# VISUAL MEDITATIONS ON THE UNIVERSE

# VISUAL MEDITATIONS ON THE UNIVERSE

JAMES S. PERKINS

*This publication made possible with the assistance of the Kern Foundation*

Quest

THE THEOSOPHICAL PUBLISHING HOUSE
Wheaton, Ill. U.S.A./Madras, India/London, England

First edition, 1984.

For additional information write to:
The Theosophical Publishing House
306 West Geneva Road
Wheaton, Illinois 60189

Library of Congress Cataloging in Publication Data

Perkins, James S. (James Scudday), 1899-
Visual meditations on the universe.

1. Perkins, James S. (James Scudday), 1899- 2. Cosmology in art. 3. Theosophy. I. Title.
ND237.P364A4 1984 759.13 83-40233
ISBN 0-8356-0233-8

Printed in the United States of America

# Contents

Seek the timeless Space within
and there in restful Selfhood
contemplate the harmony of a
primeval knowing of the universe
and its worlds and their
sublime Progenitors.

# Introduction

When Michelangelo carved his statue of David from a fourteen-foot block of marble, the stone image was not the full creative experience. The supreme rapture occurred in the beginning, with the envisioning of David created in a medium of stone sculpture. And all perceptive viewers of the image recapture in some measure the wonder of that creative experience.

There is a key here to exploring the great mystery before us—the origin and meaning of life. One may observe a parallel between Michelangelo's creating in marble the image of an archetype, and the Supreme Creator's evolving in space-time-matter the image of an archetypal universe. The mere physical formations could not be the quintessential glory of creation. The ultimate must include the prototypal vision of a Universal-Man-God perfection to be wrought in materiality.

An envisioning of that unknown splendor is possible only through cultivating our intuitive processes. A loving use of this inner faculty when awakened gives us a new dimension of creative insight into all that we seriously contemplate. Meditative explorations of knowledge about the universe, motivated by a compassionate concern for humanity's welfare and ultimate destiny, add purity and clarity of visualization to more abstract thought about the subject.

Anyone who on a clear night gazes at the starlit skies and observes the trailing veil of stars known as the Milky Way, wondering as did our ancestors of old what this huge spectacle portends, must find it useful to know that the phenomenon simply presents an edgewise view of the saucer-like form of thousands of millions of stars that are assembled in our galaxy. So enormous is this formation that its distances can be measured only in terms of *light years* (light travels at 186,000 miles a second). Imagine a journey across such an immensity! If one started this morning across the central point at the speed of light, the opposite edge would not be reached until around the year 102,000 A.D. And if a telescope were brought along to keep steadily trained on our great sun, sight of it must not be lost for an instant, else—even before reaching halfway across the galaxy—our enormous sun would never be found again. It would be as a single grain of sand on some vast ocean shore.

But think beyond our galaxy, into the space abyss. Thousands of millions of other galaxies and supergalaxies are found to exist in those unthinkable distances. The nearest neighbor to ours—the Andromeda cluster—is estimated to be 2,300,000 light-years from us. And beyond, the enormities stretch onward. Galaxies and other strange objects have been discovered that are estimated to be at 10 and even 18 billion light-years distant in the utter darkness of the abyss. Profound meditation on these awesome observations, with their mind-staggering magnitudes, leads to the realization that the normal faculties of Man are simply unable to conceive a primordial origin of the totality that includes the individual human being's existence in any significant relationship at all.

Yet intuitive meditation refutes this idea of Man's insignificance in the scheme of things. In contradiction, it suggests that *essential* Man's spiritual stature grows and expands—not diminishes—with increasing knowledge. Are the intuitive perceptions irreconcilable with the intellectual facts? The answer lies in a new comprehension of the nature of space and the formations in its embrace.

In the 1880s certain adept occultists caused the release of a wave of enlightenment regarding universal causation and the cyclic involution of Life into forms and evolution out of them. The adepts explained that what is evolving are units of consciousness unfolding in a cyclic time order through the kingdoms of Nature, to the human stage and beyond to godhood. A rich mine of information about these subjects is available in *The Secret Doctrine,* authored by Helena P. Blavatsky and published in 1888 by the Theosophical Society. It opens a new conception of Space as a continuum of consciousness, and Time as a measurement of consciousness. An even more illuminating idea is that the solar system is a *Solar Universe*! This implies that the cosmos to be explored, understood and mastered by Man, as his evolution proceeds, has necessarily to be confined to the system in which he lives and moves and has his being, his own universe amidst the innumerable universes beyond.

Another conception that arises from *The Secret Doctrine* is that the Solar Universe is spherical, and is related to the total universe spherically. A search for our primordial origins, therefore, need not extend into those light-year distances and giganticisms that modern investigations are unveiling. By exploring the Solar Universe, not only as the astronomers do, but as do the occultists—that is to say, *experientially*—universal origins and meaning will be uncovered. An intelligent acceptance of this confinement of search is, in fact, the first step in a genuine occult exploration of Life, Man, and the Universe. The great realities that concern our human stage of evolution are bounded by dimensions that are suited to Man's limited faculties and frailties. This fundamental change of outlook may be glimpsed by a simple exercise.

Suppose you could project yourself into outer space, far beyond the Earth—even beyond the Solar Universe, in deep space—and there, resting in a bodiless and timeless state of being, could experience the ceaseless universal motion and the directionless darkness of outer space. You would be profoundly disturbed by the fact that no fixed points exist from which to judge bearings, or when and how to proceed anywhere. Imagine the relief it would bring to discover that there is an ever-present cosmic direction from wherever you are that indicates the main highway of the flow of Life and all processes of creation. Like some cosmic polar current, it streams forever and everywhere in its two sustained directions: from within outwards, and from without inward to the Source of the cosmic rhythm. This flow affords, for the wise, eternal bearings upon the imperishable Center of universal being; it affords a true sense of direction for proceeding onwards.

Exploration in this direction confronts one immediately with the strange occult nature of *Space.* For Space, itself, exists in some otherness of being prior to its presence in the physical world as the emptiness medium that is familiar to us. Try meditating upon the question: Would space be here if everything else disappeared; if so, what would be its nature and condition?

More light is shed upon this obscure subject if one entertains from *The Secret Doctrine* the assumption that "Space is seven-skinned." The "skins" are shown to be energy-levels of cosmic force. These are recognized as the "planes of Nature." They are set apart from one another by the changing intensities in their atomic structures. The seven levels extend from the first, or highest intensity, to the seventh, the densest, physical level. If one adopts the nomenclature given to these planes, looking upward from the physical, the seventh, the next level beyond is the *astral* plane, the sixth. Beyond it is the *mental,* the fifth plane. The fourth level, or *Buddhic* plane, is the midlevel between the first and seventh planes. In it occurs a transformation of great significance. The third level is the *Atmic* plane. Beyond are the abstruse spiritual planes, the second being *Anupadaka,* meaning 'parentless'; and the first termed *Adi,* the primeval plane.

In the manifesting process, Life appears first on the Adi level—the spiritually highest—and proceeds "densewardly" to the seventh, the physical world. It is understood that the higher, subtler energies permeate the denser levels below. Therefore Adi—the highest level, that of the Solar Logos, the great Source of all—is not remote from the physical plane but is as closely identified with it as with all others.

With this conception of the universe as emerging from within, the electrifying intuition occurs that the space medium itself is emerging from a prior condition. This follows from conceiving the first manifestation of the universe as omnipresent Light which, later on, becomes Divine Ideation (the universe as a thought-form); and finally, at the densest level, the physical universe, with its multiplicity of disparate objects set apart and held in place by the space medium. Thus one finds that the Supreme Consciousness, descending as cosmic energy through levels of matter, characterizes the nature of each level in turn, and emerges in the densest level as the empty void that holds Earth here, Mars there, and Jupiter yonder—or, locally, you here, the tree there, and the mountain yonder. Simultaneously, the embrace of this Supreme Consciousness holds everything together—the ordered world of planets in the solar system; the elementary particles in the atom; and the infinite hosts of entities that function as a unit in the physical bodies of all creatures.

Behold, then, the glorious truth that human consciousness has the potential of rising out of all confining time and material conditions, and, through evolution, of uniting with the whole in a Union of Perfection that is centered forever in the imperishable Source of the cosmos, the supreme Space-conscious Intelligence.

The illustrations and accompanying texts of this book summarize a lifetime of studies, meditations, and visualizations grounded in a deepening awareness of the occult nature of the universe and Man's place in it. The explorations have been as voyages beyond worlds of the Known, across all horizons of physical space and time in a persevering and undismaying search for Life's origin and human meaning.

# 1

## Space

Before ever the universes are formed or the Earth and worlds brought forth, Space the Supreme Source of all exists from everlasting to everlasting.

MEDITATION: Space, the One Reality that is, was, and is ever being and becoming. . . the incomprehensible Deity whose robes are the mystic root of all matter and universes. . . both the limitless void and conditioned fullness. . . is uninfluenced by either the presence or absence of objective universes.

*The Secret Doctrine*

VISUALIZATION: A Supreme Space-Consciousness, the eternal and absolute cause of All, is perceived to be periodically manifesting universes in seven concentric spheres, or energy-levels, that extend from the omnipresent, boundless Light to the ultimate confinement, the dense physical sphere. In this sphere, Life is arriving in manifestation from both directions, the outer infinitudes and the inner nuclear space. From both, as envisioned here, Life is flowing through the energy-levels to become all that is in the physical sphere. Thus we envision Life's arrival in the physical universe as a thin, evanescent, spherical surface, denoted here by the thin black circle. This most confined space contains all the objects that we see—worlds, star systems, constellations, and galaxies—everything observable. All of these forms are composed of atoms at the cores of which nuclear energies, both positive and negative, are appearing from higher levels and disappearing back into them at enormous velocities. Yet the performance presents to our slowed-down sensory system the appearance of physical solidity, hence material reality.

The ultimate materialization that we see and touch rests on a composition of energies whose waves produce the transient illusion of objectivity. This veil of unreality is ceaselessly changing and passing away, while the indestructible Reality ever remains.

With a meditative study of the accompanying illustration—realizing that the manifesting universe is breathing forth from the mystery of the Supreme Source, and disappearing again with the inbreathing—one may grasp intuitively that the total universe is in a steady state of continuous creation. Further, every living particle is faithfully reflecting the whole. The mind-staggering totality is not perceived by the intellect but by uniting with the whole in a profound meditation that attains the beautiful simplicity of oneness, as in the simple union with a flower through its beauty, or through loving union with the flight of a bird across the lonely sky at nightfall. Through the innocent simplicity of Love one enters the Supreme Space-Consciousness that unfolds and sustains All.

# 2

## Kosmos

All beings arise from Space, and into Space
they return. Space is indeed their beginning
and Space is their final end.

*Chandogya Upanishad*

# 3

# The Solar Universe Is Born

This universe was foreordained by the eternal and changeless Law that causes the periods of rest and activity—the "Days and Nights of Brahma."

MEDITATION: Long before any traces of our universe's physical manifestation appeared, "Darkness alone filled the Boundless All. Alone the Eternal Parent" of Kosmos remained in the "infinite bosom of Duration." Father-Mother-Son were once more united—with further universes concealed in the Divine bosom. When the last vibration of the life cycle of an old universe ends and the new Day arrives, Dawn thrills through the infinitude, awakening a new wheel. A solitary Ray shoots into the shoreless sea of matter and a star is born. Within its boundary, Light blazes omnipotently, rolling back the walls of darkness. "Bright Space, son of Dark Space"—the new universe—emerges as an explosion of Light at the Adi level.

Excerpts from *The Secret Doctrine*

VISUALIZATION: The Unknowable One, termed in Sanskrit *Parabrahman*, is seen here as cold flame wrapped in profoundest mystery, pointed downward into the ocean of matter, the mother of universes. Then "Father-Mother spin a web" that embraces the new universal boundary. The solitary ray is Divine energy that brings to birth a star in its appointed center. The star is the embodiment of the Solar Logos. His light expands through the luminous sphere. It is the first veil of spirit-matter over the primordial Space-Consciousness of Parabrahman. The first veil is the solar Adi, the primeval plane of Nature that measures the circumference of the new universe. Other universes exist in untold numbers and sizes.

Is there, then, a GOD beyond God?

Astronomy provides indubitable clues. Our galaxy of stars is a vast "island universe" composed of billions of sun-stars. Each, like ours, is a universe emanating from its own creator. What, then, must be the Intelligence and power of That which is seated at the glowing center of our galaxy? Surely it is very GOD of the hosts of gods, He who sustains and wheels the mighty formation upon its orderly course. This mysterious Being, God of the galactic Kosmos, is far mightier than any of the solar creators included in the galaxy family.

Profoundly moving is the vision that relates the awesome mystery of Parabrahman with Him who is the Logos of our solar universe. For the solar cosmic order is but one of many in the greater order of the Kosmos.

Again, studying the visualization, one may discern along the frontier regions of the radiant Light a circumference that dimly but definitely marks the boundary of our solar universe. Within it the whole of solar creation is to take place. What might be its dimensions?

# 4

## A Ring Pass-Not

When the One becomes Two the Sacred Trinity appears and the Solar Universe stands measured for the eternal unfoldment.

MEDITATION: It is said that beyond the veilings of time and matter the Solar Universe is immeasurable. Yet its physical manifestation is not limitless. A circumference definitely bounds it. This is the so-called *Ring Pass-not.* It is precisely determined by three major attributes of the triune One. These are His divine powers of omnipotence, omniscience, and omnipresence.

VISUALIZATION: These measures are symbolized and visualized here as a triangle of three equal sides. At its center the Star of the Logos is radiating the three legs of the triangle, each one radiant in a tone of purest color. Each color characterizes one of the three major attributes of the Logos: the royal purple for omnipotent Will to create a universe; the sacred blue for the omniscient Love that fills and sustains the universe; and the fiery red, the omnipresent Divine Intelligence that varies the creation infinitely within the lawful Order. Upon these impregnable foundations the luminous sphere of manifestation stands enduringly.

An equilateral triangle circumscribed by a circle—this is the basic geometry of the universe. Although this symbol points to abstract values that can be realized with exactitude only in the highest levels of consciousness, intuitive meditation opens an awareness of relative physical values. While the ultimate size of the Solar Universe appears remote and immeasurable, its physical manifestation does have a definite boundary, the diameter of which may be sought by spiritual rapport and perception.

# 5

# The One Becomes Many

There is a river whose streams make glad the city of God.

Psalm 46

MEDITATION: Absolute Light is the Logos manifesting in His first aspect. The One Flaming Light shines forth as myriad streams of diamond points. These multitudinous sparks, as numberless as the stars in the firmament, are units of consciousness of the Supreme One. They are the Monads—unborn, unparented originals—the first life to appear in the universe. Joyously singing first architects, they convey the primeval Vision of the One Will unto the Designers of the Great Plan. Envision them in the splendor of Anupadaka, the glad city of God.

VISUALIZATION: The river of Life pours forth from the One Source along seven individual streams or rays. The first ray, shining outward from the Center, channels the power aspect of the First Logos. Each of the six other rays manifests a major quality of Deity. The Monads emerge in streams, each upon its own ray. The ray potentials, when evolved to perfection, will constitute in their harmonious union of differences the full achievement of the Divine plan.

Thus from the beginning the Divine potential has existed as an innate reality in every human being, as surely as does the giant oak tree lie sleeping in the tiny acorn. The overshadowing influence of the Monads indirectly guides the evolving units of consciousness in all the kingdoms of Nature.

# 6

# An Archetypal Universe

Side by side exist the Authentic All and its counterpart, the visible universe.

Plotinus

MEDITATION: With the third step down into material manifestation, the Logos unveils the Archetypal Universe. Here is revealed the supreme vision of Life, perfected in every particle, in a harmony of the whole at the end of time. "Behold Him lifting the veil, and unfurling it from East to West. He shuts out the Above, and leaves the Below to be seen as the great illusion" (*The Secret Doctrine*).

VISUALIZATION: Again we see the Star of the Solar Logos radiating the universe. The Light of Adi is establishing the boundary of what is to be. The fiery sea of Anupadaka containing the Monads is symbolized by the blue circle filled with sparkling points. The horizontal element composed of small scrolls represents the Archetypal Universe "unfurling from East to West." High in the Atmic plane the Radiant One establishes the centers against which nothing can prevail. From these eternal focuses the Solar Universe unfolds.

With the projection of this Authentic Universe in embryo, the direct action of the Logos in the evolutionary process ceases. The ultimate archetypes therefore cannot be thought-forms; they are mighty Will-centers at the Atmic level; force-vortices that are weaving glorious geometric patterns that become the matrices for the forms evolving in the lower planes of matter. Each matrix exerts a natural lawfulness of being that guides, and ultimately prevails, in the evolving forms.

How wondrous are the archetypes! Each is as some stupendous aurora borealis of glowing colors that somehow echoes hauntingly beautiful music, as though the voice of God is singing eternally in all things. The abstract Creation is formless yet radiant with an infinitude of potential forms—all that may be possible in the evolving universe. An archetype's basic vibration flashes forth its potential rhythms, and the Sea of Life responds with glories of sound, color, and form. The Ideal flows from within, and Nature's response, its path of holiness, is the evolving of form in resonance with its archetype.

# 7

## The Monad

Having brought the mind to rest in the true
Self, perceive its indestructible glory.

Shankaracharya

MEDITATION: Dwelling amidst the unparalleled splendor of the envisioned universe is the divine *idea* of Man. Named the Monad, he is seen functioning as part of the cosmic rhythm of the Logos. Conceive of this glorious being facing outward from the Divine Source to perform the promethean task of transmitting Cosmic Fire into the archetypal Self. Then he faces inward in a rhythm of return, never departing the Divine Presence.

VISUALIZATION: What mysteries veil the upper levels of our consciousness potential! From such heights, a lofty temple is Man, reaching even beyond the stars.

The Monadic rhythm resonates to the cosmic outbreathing-inbreathing life of God. Here below, the cosmic rhythm is everywhere perceived in Nature. It is evident in our necessity to breathe, the mystery of this compulsion being far more profound than the mere need for oxygen. The cycles of the heart and those of the glands and blood are further evidence of the cosmic rhythm.

The outbreathing-inbreathing rhythm is an ultimate law of manifestation. We are never outside its pulse. But much of its beneficent power is obscured by conflicts in our personal consciousness. In us disharmony too often fragments and scatters the radiant flow. Yet all natural creatures live and unfurl their potential attuned with this singing harmony that issues initially through the Monad.

# 8

## Archetypal Self

In beings in which the plenitude of these powers is manifested, there clearly the Authentic Self is present.

Plotinus

MEDITATION: The Monad shines forth in the Archetypal Universe—an individualized paradigm of the perfect Man-God to be. This is the eternal *Monad turned outward* upon its particular ray, ready to unfold in the new universe. This archetypal Self is the "Heavenly Man," for he is literally the father of all that the human being is and ever will be. Through that Self the rhythmic flow of divine beneficence is channeled. We receive it as the "daily bread" of freshened spirit and renewed strength. With the dawning of each day we breathe in its nectar.

VISUALIZATION: In loftiest majestic form is glimpsed the archetypal Self as symbolically poised upon a sphere of matter, in triumphant mastery of the phenomenal worlds of illusion. Above is the star of the eternal Monad, raying down the Divine Life into and through the archetypal Self to worlds below. The winged visualization suggests a sphere of radiance amidst which stands the god-like figure of perfected Man.

Throughout all the deaths and rebirths to come and all the perils, disasters and triumphs to be experienced, the consummate victory of the Divine Self is assured—and is ever reassured—by the rhythmic cosmic flow from the eternal Monad. We commonly experience this inner renewal and recurrence, though its source is little recognized.

But the wise ones turn instinctively to hail the god within as the shield and source of strength, affirming the unity of self with higher Self from the beginning.

# 9

# The Mighty Pilgrimage

Let us then withdraw from the world-image with its compelling illusions and enter that world of Reality where alone living Truth can be experienced.

J.J. Van Der Leeuw

MEDITATION: The myriad Sparks of the One Flame, the Monads, are linked into the archetypal world. Each Monad, now an archetypal Self, an individual god, prepares for the mighty Pilgrimage of Awakening. What does each behold gazing across the eternities before even the planetary rings are formed? What awesome Design is unveiled?

VISUALIZATION: The intention of Life's total process and purpose is revealed in a single sublime panorama. In the Above, the upper region, are the three abstract stages of direct Logoic creation. The Solar Logos (shown as the star at the center) radiates the Light measuring the limits of the universe. Hosts of Monads in Anupadaka appear as sparkling points on a blue field. The archetypal universe is manifested by the third aspect of the Logos. It is located in the upper strata of the Atmic field. These three fields of Divine energies are the Above. They are mirrored in the Below, as the five-fold levels of the Shadow Universe.

Behold then, the obligatory Pilgrimage: all archetypal units are drawn into the cosmic tide (shown as green arrows) streaming into ever-denser levels of matter, with experience at each level fully recorded in the permanent atoms (the red circles). The tidal flow in succeeding waves is relentlessly downward into confining materiality. It is the great outbreathing of the Logoic Purpose. As the descent of each wave reaches its deepest penetration into physical matter, the long outbreathing motion pauses, then slowly is reversed. The cosmic inbreathing commences. With it, Life's unfoldment begins for the units of consciousness. They live and grow through the kingdoms of Nature, mastering the realms through which they came: mineral, vegetable, animal, human, and divine. As the forms become more refined and sensitive, the Monads unfold more swiftly. In time the evolving waves advance to the level of humanhood.

At this point comes a mighty leap forward: each unit "individualizes" from animal to human consciousness. The upthrust animal consciousness (symbolized by the green triangle) has attained an almost human stage through the upthrust, thus calling down the flaming golden triangle of the Monadic link. The two triangles merge: this is the glorious transcendence of the animal consciousness lifted into the human stage. Experience through aeons in the lower realms of Nature now becomes memory buried in the subconsciousness in Man. It is joined with the Monadic potential in the superconsciousness.

The god-Self unawakened—sent forth upon the journey into densest levels of matter—has now attained an individual egoic selfhood, an immortal soul. Through the long cycles of reincarnation, the soul's vehicle grows clearer and ever more wealth-filled. At last full spiritual awakening dawns and ultimately complete liberation of the god within.

# 10

## A Cosmorama

The one Life-principle when in action runs in circuits; and so with the formation of worlds and in the great descending and ascending circle of necessity.

*Mahatma Letters*

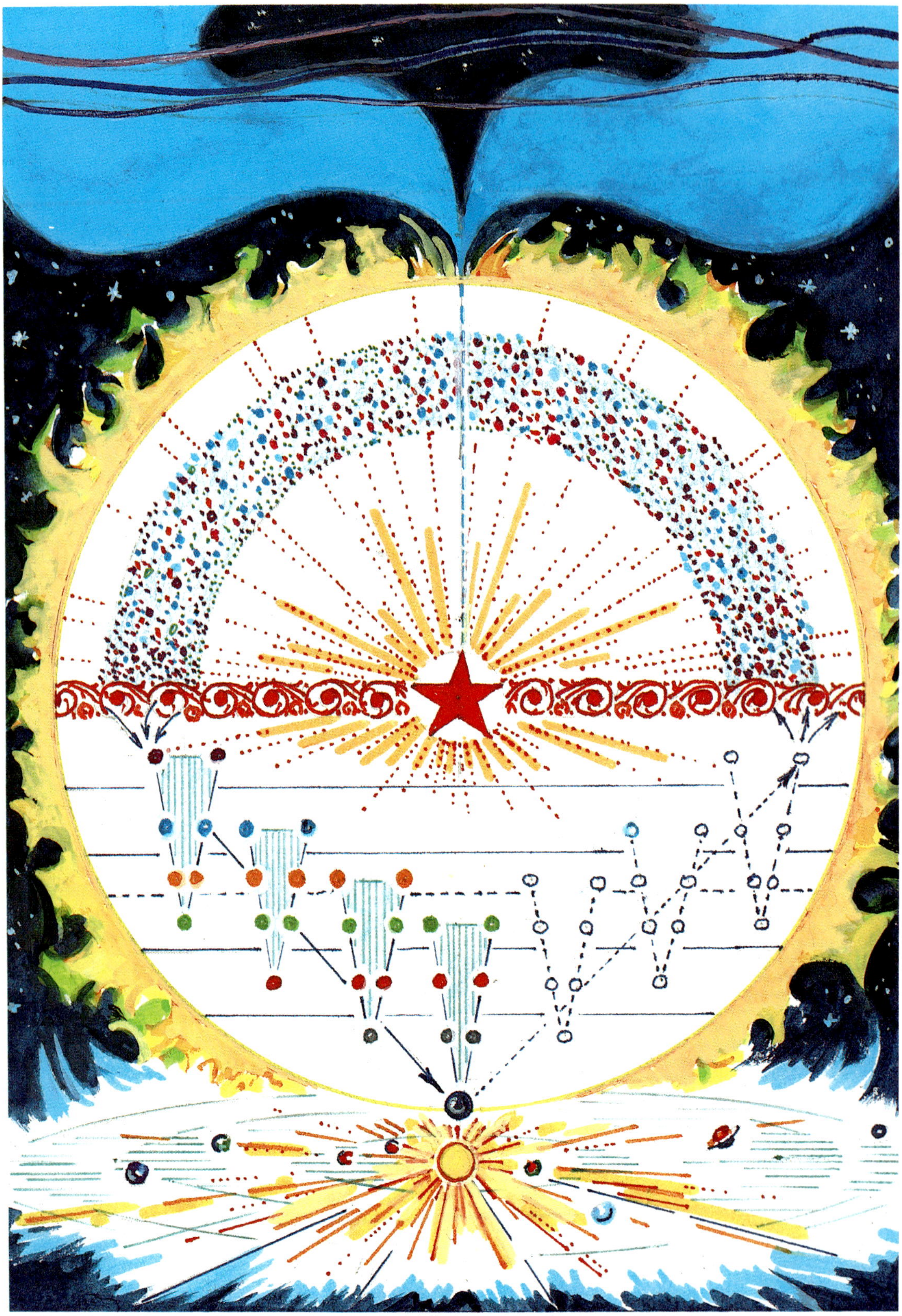

MEDITATION: By what cosmic mechanism does the great "cycle of necessity" flow with orderly precision into and out of the worlds of form?

VISUALIZATION: The Life wave proceeds very slowly through enormous periods of time. *Chains* of globes provide the arena for Life's elliptical motion into matter and back again. Each chain consists of seven globes. Four of these step Life down into densening matter. At the midpoint the cosmic tide turns upward from descent into material dominance to the liberating of consciousness. The movement upward then slowly rises through the three remaining globes of the chain, back to the original level. Hence in a chain there are three globes for the involvement in matter, a fourth for the struggle of disenthrallment and victorious reversal of direction; and three globes of return, reviewing and consolidating on all planes the results of the downward experience.

A brief study of the visual "cosmorama" will clarify many questions. At the bottom of the design appears the visible sun and orbiting it the nine known planets in the flat pancake arrangement that represents the astronomical solar system. Above are depicted the invisible chains of globes that belong to the Earth Scheme alone. Imagine all the other invisible planetary schemes of chains and globes that are evolving in our solar system. One can see how the astronomical solar system is in reality an occult *Solar Universe.*

In our Earth Scheme, the earliest globe at the Atmic level receives the Life wave directly from the archetypal world. The wave descends through the first four globes to the mental level where the first actual forms are created, thought-forms. Then the Life wave returns step by step to the Atmic level via the three other globes of that first chain. The Life wave circles seven times around the chain. On the seventh round the globes die out one by one and vanish forever as Life departs from each.

The second chain is then born and develops. It is in certain respects a reincarnation of the first chain. After it follows the third chain. Its fourth globe, still disintegrating, remains as the mystic satellite of our Earth—its moon.

At present our Life wave is in its fourth chain, upon its fourth round of the chain, well beyond its midpoint. The midpoint reversal of direction took place in the dim past of this planet's life. The tide of evolution on Earth is now on its way upward, with the evolutionary purpose of refining all physical forms for embodying Monadic life more fully. The large-scale direction is toward a united and cooperative humanity, in harmony with all the kingdoms of Nature.

Looking far ahead, there are three more rounds of our Earth chain, and beyond that, three remaining chains. These lie in the distant future, when mankind will see stupendous civilizations evolved by superhumanities that will bring our wave of evolution to its destined stage of perfection.

These chains are not strings of globes geographically separated in space. More realistically conceived, the globes are concentric with a common center, each contained within the other. In the fourth chain, for example, the globe of finer mental matter contains the denser astral one; and it, in turn, the etheric globe. The innermost, the physical Earth, is the central core of the chain; and this one *is* placed geographically with relation to other physical globes, such as Jupiter, Venus, etc., that are in their own chains.

All movements downward and upward are negotiated through vibratory changes in the states of consciousness of whatever is in motion. A person or planet remains where—and as—they are until changes in their vibratory rate unveil for them locations elsewhere.

So colossal is the splendor of cosmic order that we must dare wondrously if we are to capture the whole of it in a single glance and grasp its awesome totality in our frail barque of language.

Let us now follow more closely our particular Life wave's evolution through the Earth chain alone.

# 11

## Chain One

Life is a globe of Light that crystallizes into a mineral; the mineral becomes a plant; the plant a tree; the tree an animal; and the animal metamorphoses through Love or anti-love into Man.

MEDITATION: What can be known of Life in the beginning—in the first chain? It must have begun in a blaze of glory unsurpassable, for the Life wave was only one step away from the sublime archetypal world. Each unit of consciousness, born of its Monadic progenitor, was still closely linked to the divine source at the commencement of the mighty "cycle of necessity," the pilgrimage into the planes of matter. The first step down into the "mineral" state of awakening was experienced on the very first globe, at the *Atmic* level—in Nirvana! At that level everything exists as some point of consciousness in the union of the whole.

Somewhere in the profoundest depths of our subconscious selves exist layers of memory of that stupendous period in our cosmic past. Glimpses of its essentials may be awakened.

VISUALIZATION: In envisioning this utterly remote and wondrous circumstance, we begin with the star at the real center of the scene. Let it now represent the archetypal Selfhood of any single Monad. As the Life wave begins its descent, the irresistible tidal flow carries with it all units of consciousness that are destined to unfold their Divine powers in the seven chains of the Earth Scheme of evolution.

Globe A is of an Atmic nature. Everything there is totally identified with the One Life, in a state of non-self-conscious Divine bliss. This first globe cannot rightly be denoted as having a global form because no forms, as we conceive them to be, exist at that level. Globe A is more realistically conceived as a radiance of forces shining universally from a cosmic center. The units of Life are occupied with evolving Atmic vestures. Through these they are exerting primordial impulses that issue directly from the mineral archetypes created by the Monad's obeying the Will of the Logos. Every impact and gradation of change in globe A registers some mode of the blazing glory.

Through untold aeons of time the mineral archetypes of each of the seven rays are thus transmitted into the Atmic levels. The enormous pressure of the divine Will impregnates the units with their specific ray qualities for the eternities that lie ahead. Each unit of Life summarizes and records all experiences in a single Atmic center known as the *Atmic permanent atom.* This computerizing operation serves in the creation of one of the basic elements of the inner structure of the evolving unit of consciousness.

From Globe A the Life wave moves downward to Globe B, which is centered in the Buddhic plane. The chain is characterized by the new factor: *relatedness.* There are definite distinctions now appearing between the more and the less attractive assemblages of unity. Nothing yet exists that is not some expression of divine Love. No separateness appears anywhere. The units are attracting ever-renewing relationships within Life's total unity. The mission of the Monads is to employ their units of consciousness to radiate the mineral archetypes, clothing them now in Buddhic matter. Again, every experience is recorded by each unit in a *Buddhic permanent atom.* (Visualization 11 depicts this growing apparatus of permanent atoms, placing it directly under the central star.) Through the permanent atoms the Monad is assembling an enduring spiritual vehicle through which to operate effectively in lower worlds of matter.

From Globe B the Life wave descends to Globe C, of a higher mental state denoted as *Manas*. Here the units experience stronger vibrations of group attraction or repulsion. But all experiences are, as yet, in "nonobjective" matter. A third element, the *Manasic permanent atom,* is now added to the Atmic-Buddhic string. This Atma-Buddhi-Manas complex when completed provides the Monad with an immortal vehicle for channeling spiritual forces into the evolving units below. When the units, with their newly established atomic complexes, began to be associated more closely in groups of like kind, the creation of "group-souls" occurred. The group-soul mechanism affords each Monad with a far more extensive means for experiencing Life's variety on globe D, in the deepening levels of matter of descending chains.

With completion of the seven rounds of Chain One, the units of consciousness succeeded in bringing down authentic replicas of the Divine mineral Archetypes into the form-building levels of the Mental plane. There they stand, established for futher mineral evolution. In the right hand panel of Visual 11 are shown the permanent atoms added to the units of consciousness in Chain One.

# 12

## Mineral Archetypes

A celestial paradise of flashing color—of Monads descending and evolving milliards of jewel forms in the bliss of archetypal birth in Nirvana.

MEDITATION: At the nadir of the first chain—in mental matter—the first forms take shape. The mental globe in all its iridescent beauty is a teeming world of glistening formations. The seven rays of mineral archetypes break into a glorious diversity of mental structures.

VISUALIZATION: We are following the evolution of one Life wave only, through the chain system. In the first chain, that Life wave evolved only the mineral kingdom. However, we must not overlook the fact that at the same time all of the kingdoms are present in other Life waves. The waves of Life are continuous; there is no break in their rhythmic manifestation. Amidst the multiplicity of evolving entities, we are pursuing developments of the units of consciousness in our Life wave which later on in Chain Four will evolve the present humanity on Earth. In Chain One the mineral kingdom is evolving the earliest stages in the *making of Man.*

The ray influence is sharply marked in the mineral kingdom, even at the mental level. Diamonds are evolved by first ray Monads; sapphires, by second ray; emeralds, by third ray; and so on. Imagine the ballet splendor—the lawful precision—with which the star Monads translate the Divine Archetypes from their primeval glory into mental replicas! The Archetypes are ideal jewel forms of what is to be evolved in future chains. The visual design suggests the full ordered procedure by which the Monads evolve from the waters of Life the growing rock and crystal formations that ultimately achieve the perfect reflection in the mental level of the Divine Archetypes for the mineral kingdom.

# 13

## Chain Two

Life's plan glimpsed at first in faintest outline
gathering on one's mental horizon unveils
with astonishing lucidity the perfect Design.

MEDITATION: Chain Two awakens an entirely new state of freedom, another world of experience. The units of consciousness perform an enormous intra-kingdom leap ahead. From the constrictions of mineral existence, they advance into the plant kingdom of movement. What does this mean for their growth?.

VISUALIZATION: After the seventh round of Chain One the Life wave circled on, leaving behind the dying globes. With the final departure of all energies, the globes, one by one, disappeared forever. Then followed a period when Life in all the kingdoms reposed in deep sleep for a duration equal to that of the life of the chain.

As everywhere in the universe, here too birth follows death in cyclic progression. In the dawn of its awakening, Chain Two began with a radiant new sphere of Buddhic light, a step below its Atmic progenitor of the former chain. As the Life wave rolled on in its endless cycles, one after another of the globes of Chain Two were formed, each one denser than its predecessor, reaching at last the Astral stage of matter. In that heavier materiality the living units were no longer merged and intermingled in close identity with each other. Amidst the strong astral currents of repulsion and attraction, the units experienced for the first time a definite separation from one another. Distinctions of character and differences of direction were accented. The growing divisiveness foretold the struggles in later times of Life's evolution in utmost variety of individual forms. In the visualization, note that the soul's vehicle that is being formed of permanent atoms now includes an astral center.

# 14

## Vegetation Archetypes

Before the advancing units of consciousness can be suitably embodied, a vast preparation of forms in the various fields of evolution must take place.

MEDITATION: New freedom of movement accompanies the Life wave's penetration into Chain Two's denser fields of matter. As the Monads transmit reflections of the Divine Archetypes to the chain's globe D, an astonishing variation of forms appears.

VISUALIZATION: The units of consciousness entered Chain Two's first globe of Buddhic matter in group-soul assemblages for the plant kingdom evolution. They were experiencing a renewal of the bliss of total unity at that level. Globe B of Chain Two was of higher mental material, a brilliant world of variations of the plant kingdom. It opened the dimension of self-movement, which developed further in Globe C. Globe D of heavier astral matter with powerfully weaving forces of attraction and repulsion ceaselessly divided the evolving forms, increasing their struggle for existence. Stronger forms simply absorbed weaker ones. This demand to develop more effective vehicles actually augmented the archetypal influence upon the developing forms. The environment compelled greater excellence of performance.

The ray differentiations in the plant kingdom became far more distinctive and discernible on Globe D as evolution proceeded through the seven rounds of Chain Two. Visual #14 suggests the ordered transmission of plant kingdom archetypes into Globe D of Chain Two, the astral planet. The Monads are seen radiating the ladder of archetypes from grasses to flowers, shrubs and trees, to the archetypal kings of the plant evolution that will be achieved in future worlds.

# 15

## Chain Three

Human thought can hardly envision the Monad's resistless rise through the cycles of life and death in many forms, on globes and cosmic chains to the triumphant birth of Man.

MEDITATION: While Chain Three advances the Life wave deeper into the folds of matter, the evolving units of consciousness are progressing steadily upward toward mastering the conditions of limitation. The Monadic link of permanent atoms now extends to the etheric level of the physical plane.

VISUALIZATION: After the long interchain rest period, the units awaken in Chain Three to a still more glorious freedom of volition in a higher order of life's forms. They have leaped forward from the plant kingdom into the animal kingdom and are opening a wholly new dimension of existence.

Globe A of Chain Three, in higher mental matter, was the last globe of the descending chains remaining fluid enough to reflect directly the Divine Archetypes in their exalted level. Globes B and C descended into the heavier materiality of the lower mental and astral matter. Globe D penetrated the densest level of all—the physical—in its subtler regions denoted as *etheric.* The Life wave's sojourn in this superphysical world has extraordinary significance for mankind, for it is the period immediately preceding entrance into the fourth chain's evolution of humanity. The inner structure of principles now includes an etheric center. The panel depicts the additional ultimate permanent atom in dense matter—the etheric (subtle physical).

# 16

## Animal Archetypes

Never does the eye behold Archetypal Beauty in Nature's forms till the Soul itself be beautiful.

MEDITATION: In the beauty of wind-whipped clouds in a golden sunset may be glimpsed forms of such grace and symmetry as the Divine Archetypes' mandate for the animal kingdom.

VISUALIZATION: Evolution of the animal kingdom in Chain Three resulted from transmission of the archetypes into beautiful forms on Globe D. These forms are the prototypes of animals that will become fully evolved physically on Earth in the Fourth Round. The prototypes presage the exquisite skill and grace of action that the animals experience in their new dimension of freedom. Life for them in their natural environment calls forth the potentials of their archetypes in rhythms of animal ecstasies.

# 17

## Group-Soul Evolution

The group-soul existences in mineral, vegetable and animal kingdoms are stages of growth leading to the complete differentiation of individual Souls in the human kingdom.

MEDITATION: A single Monad enfolds and channels its spiritual forces through any number of evolving physical forms. The Monad, while influencing the perfecting of forms, also directs evolution toward the awakening of higher consciousness.

VISUALIZATION: Animal consciousness is not only evolved through individual forms, but through group experience and growth. This group development occurs throughout the third chain, particularly during the seven periods during which the Life wave occupies its Globe D. Animals are united in groups, each animal sharing the soul life of its group. A flock of birds swings in flight as though activated by a common will. A school of fish or herd of gazelles perform as if guided by a single intelligence.

The group-soul is an admirable conveyance for sharing the collective experience of numbers of animals. A number of lions may be seen linked together in a single sphere of consciousness, as illustrated in Visual 17. Each individual animal is participating in a pool of experiences contributed by all the individuals of the group.

Through the common shelter of this immortal group-soul that is experiencing life and death in many forms, there is a constant stimulation of consciousness in all the creatures attached to the group. The sheltering feature of group-soul existence indicates that the animal is not aware of *aloneness.* It faces life from a shared—not an individual—standpoint.

In the illustration the star Monad above is radiating its influence into the sphere through the medium of an angelic presence that represents the immortal group-soul intelligence. This entity is visualized gathering the essential quality of each member's experiences and redistributing it to the whole group. Thus each lion has a "soul"—an inner guiding intelligence—but it is the same soul for all in the group.

Into this arena of mutual existence radiates the dominant ray influence from the Monad. The ray factor is a decisive feature in group-soul evolution. A first-ray Monad evolves through a first-ray group-soul; and only first-ray animals are attached to it. Lions will be found only in a first-ray group-soul. All animals are evolving their natures on one or another of the seven rays.

Thus it may be seen that there is a common influx of spiritual forces from above, in addition to the flow of impacts from all the environmental experiences pouring into the group-soul. Together these unfold the lion's characteristic nature and action. The same is true for all other animals.

There is individual progress in the group-soul. A domestic animal in a group-soul, due to human association, advances more rapidly than others of his type that remain in the wild kingdom. When this individual progress occurs, the group-soul may be reduced in number, finally with only two or three very intelligent and beautiful animals. Thus the animal unit of consciousness gradually approaches entry into the human kingdom.

# 18

## Prehuman Group on the Moon

The community consciousness, born of the group-soul evolution, will reach out in evolved mankind to include the whole of Earth's human family.

MEDITATION: The group-soul creatures enjoy a very close unity. They are totally obedient to a single instinct, acting together as by a common will.

VISUALIZATION: The great advance upward in evolution occurs invisibly, within the inner structure of the units of consciousness—whether or not there is outward resemblance to the human form. Hence animals approaching humanhood are not necessarily ape-like in appearance. The apes are on the fourth ray. Dogs, cats, horses and elephants represent other ray doorways into the human kingdom.

Envisioned here is a group of prehuman creatures in forms such as we ourselves might have required at this stage of our evolution many aeons past, when we were in the third chain on its Globe D. This is a small group, very tightly knit in their united consciousness, as represented by the sphere of light which indicates the single group-soul they share. In it are seen the Manasic-Buddhic-Atmic permanent atoms that structure and color the immortal group vehicle, an instrument of the eternal Monad to which it is linked. When the animal reaches humanhood, the group-soul, located in the higher mental level, will metamorphose into the Causal vehicle that houses individual Man's immortal soul—the vehicle for Atma-Buddhi-Manas. In the illustration the topmost figure has a more human appearance, evidently more advanced and nearing the time of individualization.

# 19

# Individualized Man

From the beings who live aloft in the heavens
Man is born.

Hermetic Fragment

MEDITATION: The central problem of human life is the element of aloneness. It is solved in part by family unity, religious unity, national and cultural unity—all leading to the mystical realization of human unity, the family of Man. Beyond is the realized unity of all Life, and the Source of Life itself, wherein aloneness vanishes forever.

VISUALIZATION: The long ages of group-soul experience lie buried deeply in humanity's subconsciousness. Traces of it sometimes surface in situations of high danger that call forth "herd instinct," as in time of war and invasion. It also leaps out when there is panic, or mob violence. More often it is experienced as that strange yearning for companionship that surfaces so strongly when one is keenly aware of the basic aloneness of the human individual confronted by life's vicissitudes and its inevitable, but uncertain, terminus.

Depicted here is a lone individualized human being facing the challenges of life in the physical world. But he has now an inner power that never existed before. The miracle has taken place in the soul. It is evidenced by the interlaced triangles in the design's background. The lower green triangle is the chalice of consciousness that has evolved in the lower kingdoms. It is held upward receiving the downpouring of the yellow triangle representing the immortal vehicle of permanent atoms, the Atma-Buddhi-Manas vessel that the Monad, the star, has assembled through the ages. The Monad's influence is now directly interwoven with the life of the individual human soul, adding to it the potentials of godhood. The barrier has been crossed that forever separates the human from the animal kingdom.

Man's individual way is ever upward with unfoldment of divine powers, to the ultimte mastery of all the forces of earth and heaven.

# 20

# Chain Four—The Human Kingdom

The Divine spirit in man cannot fall back
again, but rises ever onward and upward
toward the Divinity whence it came.
C. W. Leadbeater

MEDITATION: We now arrive at the most enthralling and illuminative period in chain history, in which our Life wave evolves the human kingdom. While the cycles of Chain Four descend to the ultimate density in the solar system—the physical plane—with all entities buried in deepest materiality, the Life wave advances upward to its peak achievment: the liberation of consciousness through the human individual. But in the fourth round of our Earth chain, and at the present time, physical life exists under three heavy blankets of matter that prevent clear perception of the sublime realities beyond. To unveil these realities requires an extraordinary exertion of higher energies. And this is the soul power to be awakened and developed in the human kingdom in the fourth chain.

VISUALIZATION: Globe A is centered in the lower levels of the mental plane where every exercise of energy is instantly enclosed in mental form. Globe B is of astral matter that seethes with vortices of attraction and repulsion. Globe C supplies the semiphysical condition for the conveyance of the Life wave into the dense physical Globe D. This transition is supplemented by temporary use of the etheric conditions on two nearby physical planets (Mars and Mercury). Globe D is the Earth planet. Here the evolving human body encases all seven principles of Man's inner structure of consciousness assembled in the preceding chains. The field is now prepared for the Self's unfoldment within its *Auric Egg* as shown.

# 21

## Birth of the Planet Earth

The Sun is the giver of Life to the whole planetary system; the Moon is the giver of Life to our globe.

*The Secret Doctrine*

MEDITATION: The densest globe in Chain Three survived long after that chain perished, as it was destined to become a physical parent of Globe D of Chain Four. How might this transition have occurred? How might the Moon have mothered the Earth?

VISUALIZATION: From the standpoint of genuine occultism, any proposals concerning major astronomical events must be in harmony with the universal principles rooted in the greater Reality from which all Life stems. This means that the birth of a planet is viewed as similar in essentials to the birth of any form of physical life, in that all require a reproductive process.

The universal pattern of generative causality in the physical world is orderly: male seed, delivery to womb, female nourishing, infant birth. The fundamental principle that governed our Earth's origin was purposeful in time and orderly in sequence—not by chaotic, fortuitous catastrophism. Planetary birth is analogous to the first appearance of any living organism in the physical scheme, that is, through a vital center-point, a seed or cell. In the Earth, as in all living forms, magnetic forces radiate from the point and attract appropriate material for the manifestation.

Regarding Earth's source, then, we may seek in the solar universe the four causal factors that afford a planetary birth in harmony with universal order. These are male and female parents, seed, and transmittal of seed. The first is obviously the Sun, the giver of Life, father and sustainer of the solar family. The second is the mother, Globe D of Chain Three. Still alive but inert, she is prepared to provide the substance for the new planet, assisted by densifying agents. As for the third generative factor, the mysterious seed to be transmitted to the Earth locality, we must search further afield. It is known that far beyond the outer regions of the solar sphere clouds of cometary seed are slowly orbiting the Sun. Here, logically, can be the seed source—the third generative factor—for which we are seeking. The transmittal of the seed to the Earth's location suggests a dramatic scene of action.

The astronomical scenario might be thus pictured: Globe D had existed for aeons as a life-bearing planet moving in its orderly path about the Sun. At the conclusion of Chain Three's life cycle, its Globe D fell into a deep state of pralaya or rest. Its sleep is now to be disturbed.

A comet appears on the cosmic horizon and is plunging toward the Sun. As it reaches perihelion, very near the Sun, it is "captured"—that is, its giant parabolic course is interrupted. The fierce solar heat has transformed its frozen nucleus into volatile gases that eject jet streams that alter the comet's course in its orbit. This solar performance changes the very nature of the comet itself.

Reborn of the Sun, so to speak, the whirling phenomenon is directed straight toward Globe D. It becomes a spinning vortex of gravitational forces, drawing into its sphere clouds of dust and solar debris from surrounding space. As it approaches Globe D, the giant fireball fills the sky with its great flames. Enormous tongues of fire envelop the hitherto comatose Globe D, melting its crust. Huge masses of the globe's surface peel off in seething cauldrons of material that stream across the gulf separating the two orbs. The immense gravitational forces from without and volcanic pressure from within Globe D cause explosions that shatter its surface on the side that faces the newly birthing planet. Some of the crustal material hurtling outward into frigid space splashes back upon the molten globe, creating vast basins ringed by mountainous waves of molten lava that freeze into place. The old planet is now becoming a scarred and emaciated orb that assumes the humble role of satellite of her own progeny. Her torn face is forever turned to the newborn Earth as a solemn reminder of the awesome event of its birth.

As a cosmic parent, she had given her substance to the delivery of a wailing new infant in the skies. No longer herself a beauteous planet orbiting her solar lord, this mother of Earth settles into a mystical path of attendance upon the new member of the solar family. In strange and haunting ways, the rhythms of all earth's creatures are touched by her silvery passage across the heavens each month.

After ages of molten tumult the new planet cooled; crustal surfaces began to appear. Earth was being prepared to receive from the dead third chain the kingdoms of Nature that would awaken the new globe to full vitality.

For aeons the Moon's very existence has been silently teaching Man that in the universal economy every new form arises out of the substance of older forms. New forms are freshly assembled by Life's orderly continuity as it passes through the chains of evolution on its return to the eternal Source of All.

# 22

# Lemuria—Human Evolution Begins

But such a tide as moving seems asleep,
  Too full for sound and foam,
When that which drew from out the boundless deep
  Turns again home.

Tennyson

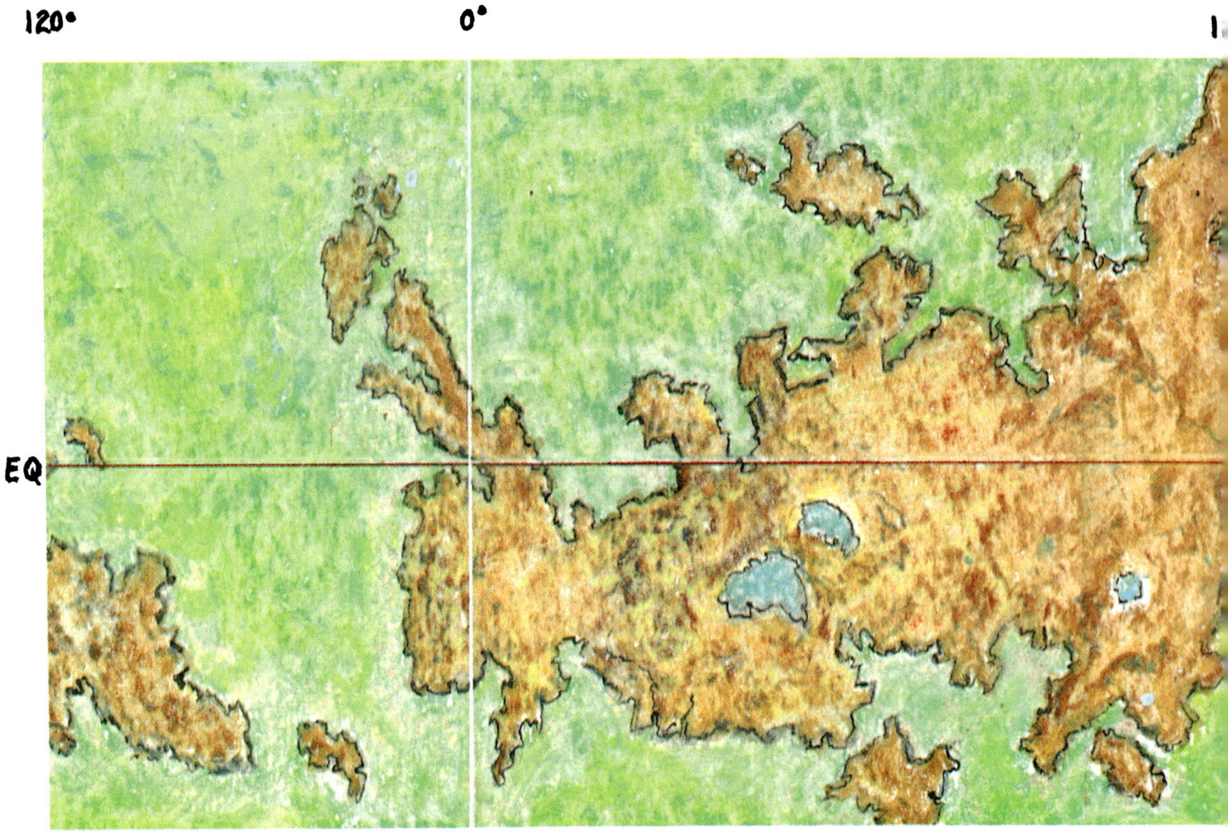
120°
0°
EQ

MEDITATION: Into the molten planet slowly congealing, the human wave began arriving. Although technically "human," these beings were neither in physical forms nor were they human in appearance. At first, all forms were of subtler astral and etheric substance, unaffected by the surrounding physical conditions. What were the units of consciousness experiencing in those strange conditions?

VISUALIZATION: A map from occult sources shows the giant land masses that were forming an enormous continent in early geologic times. Ages after the Earth was born, this vast continent spread around the southern hemisphere of the globe. Named in some quarters Lemuria, it was the land where the human race would complete its descent into densest physical materiality and begin its ascent back to the Source through the evolution of physical forms toward archetypal perfection. At first, the forms were of superphysical matter, gigantic and of misshapen appearance. Their progress was toward further densification and reduction in size. In time the bones hardened and the skeletons became enfleshed, with organs and nervous systems evolved. But with this densification, Man lost sight of the invisible worlds and their inhabitants. His perceptivity became limited to the organs of the five senses.

Concurrently with this materialization into physical form, there occurred a further cutoff from consciousness of Life's extension from the invisible worlds—the inexorable separation of the sexes. Earlier beings had no need of dual sexual reproduction. Every individual embodied both masculine and feminine forces, and reproduction took place from a single individual. But descent into physical materiality rendered necessary an evolution of the sexual means of creating new bodies. Evolving through long periods of time, dual-sexed individuals tended to become single sexed, slowly attaining either masculine or feminine bodies organically arranged for reproduction. As the separation of the sexes progressed, the hitherto whole Man became a half-being—either masculine or feminine—each half driven thereafter to seek the other for a natural completeness of selfhood. This was Man's ultimate incarceration in materiality, with but a single outlet open by which to experience vividly the joy of Life's eternal unity—through ecstatic cohabitation with the opposite sex.

Moreover, life in physical bodies became subject to the aging process, to disease, disablement, and death. Mortals also must have daily nourishment and the provision for shelter of some kind for healthy existence and propagation of their kind. Early humanity, therefore, driven by these earthly necessities, shut off from direct communication with the reality of the invisible life around them, harrassed by hungers and fears—and embattled by Life itself—tended to become violent and savage in their attempts to obtain what was needed.

Thus the tidal flow of the great Monadic pilgrimage into densest matter sank to its nadir; and the tidal reversal from involution to evolution commenced for humanity. The long struggle began between Man's enthrallment in materiality and the liberating of his consciousness from total identity with matter. A triumphant outcome must await aeons of evolution toward enlightened consciousness.

The problem now was the self-will in Man that had grown monstrously strong. Conflict was universal. Cruel evils spread with dire consequences. Again and again continental destruction was visited upon Lemuria by the karmic law. The time had arrived for aid from a "sister system." While Lemuria yet existed, the great Teachers arrived.

The Lemurian period of human evolution was crucial because of the three events of cosmic importance that took place there. These were the reversal of direction of the great cycle of involution to that of evolution for mankind; the separation of the sexes; and the arrival of the Adepts from the Venus system, bringing with them the knowledge, wisdom, and powers of a more advanced humanity.

# 23

## Arrival of the Great Teachers

One God, one law, one element
And one far-off divine event,
To which the whole creation moves.
Tennyson

MEDITATION: In the cosmic family of the Solar Universe there are "elder brothers," humanities far more advanced in evolution than other "younger brothers" in less evolved, planetary chain-systems. When needed, certain of the elders may be directed to aid the development of younger humanities in sister chains. Such is the Earth-Venus relationship.

VISUALIZATION: Although Venus is sometimes referred to as Earth's sister planet, it is infinitely older than the Earth. The Life wave that once occupied the Venus planet is now in the closing period of its fifth chain. The humanity—all now super-Adepts—graduated ages ago from all physical limitation and any need for a physical planet. Venus is now a disintegrating globe that is drawing near physical extinction.

So it was in the orderly cosmic scheme that those known as Lords of the Flame came from Venus to Earth. In the visualization they are seen descending upon a far northern coast of the Lemurian continent, where once there existed in an inlet a lovely island near the sea's southern shore. It was from this point on our planet, traditionally referred to in the East as Shamballa, that evolution on this globe was quickened through intensified governing of the forces created in Nature by its creatures. Here the center was established for the enlightenment of humanity. To it would be drawn the most advanced among men for instruction and training. Among the eternal verities to be taught were the One source of Life; the nature of Man and the universe; the hierarchies of Beings; the hidden laws and powers to be awakened in Man through a responsible obedience to universal principles and lives devoted to the welfare of the whole.

Age after age those who were thus instructed scattered their knowledge among the peoples of the world, thereby planting the seeds of a universal wisdom-religion. This spread in a variety of forms into the primitive tribal peoples and later into the more civilized cultures of the archaic ages.

In later Lemuria the human form diminished from its gigantic, inchoate appearance. Due to enlightened rulership, civilizations arose and the arts and science began to appear. But wherever the greater Teachers departed and the positions of power were occupied by less able men, the advancements quickly degenerated. In such times, cruel and selfish barbarism erupted and destroyed much of the progress that had been made. The cycles of Man's toiling way upward are tedious and painful. Disenthrallment from the grip of dense materiality and a growing mastery of matter are possible only through spiritual enlightenment and discipline.

The ultimate struggles destined to set the course of human evolution definitely upward for all future ages were yet to take place on another continent inhabited by a far more vigorous and willful race of mankind.

# 24

## Enlightenment

And when the subterranean flame shall
   burst its prison and devour the frame,
Thou shalt be still as Thou wert before.

*Rig Veda*

MEDITATION: Awesome indeed is the vision of Reality and the knowledge conveyed by the gods from Venus. Chanted in lilting verses in a language of form that is music, and music that is color, they voiced the story of the unrolling manifestation of worlds and universes, and the awakening of divinity in all creatures.

VISUALIZATION: When the Lords of the Flame came from Venus, they sounded the wondrous Logos doctrine of how cosmic ideation poured forth into cosmic substance, generating schemes of worlds, universes, gods, and mankind. They revealed the miracle of birth of the Monads and their venturing forth to unfold divine powers in the service of the Cosmic Cause; and of their return in the universal perfection achieved at the end of time when all again are enfolded in the Source of Being.

They who had themselves achieved this mighty end declared with profound assurance, that as surely as the rains of heaven descend to earth's shoreless oceans and rise again, merged in a life more wondrously free, so does the mysterious god in Man rise through the aeons of toil and growth to reach that union of the "Day-be-with-us" when all are once more gathered in the shining glory of the One Consciousness.

Man was admonished to learn of the One Life and the few essential truths to be realized in action. To abide by these is the essence of the Wisdom-Religion that in archaic ages was spread across the world. Certain of its fundamentals exist today in the great world religions and genuine occult teachings. The essential enlightenment that concerns this modern age is the knowledge of our origin, together with the inner structure of our being, and the laws governing it. With this knowledge and the efforts to apply it, the secret of our own radiant sun-Self will be revealed and its powers awakened.

The symbol visualized here sets forth the stairway mounting to these heights. It begins in the murkiness of the material planes. The foursquare foundations are seated firmly in the physical, etheric, astral, and mental bodies. These, united, compose mortal man's consciousness, whose growth is upward through the process of reincarnation.

Above is seen rising a triangle symbolizing the immortal Self, with triple powers derived from the principles of Manas, Buddhic, and Atmic consciousness. Together, these structure Man's spiritual soul—his deathless Selfhood, pictured here as a victorious figure winging its way upward to union with the Divine Self, the Star-Monad. Beyond, in the Himalayan silence, is That of which nothing can be said, for it concerns the individual's ultimate flight alone into the Alone—in the boundless, infinite and eternal bliss.

# 25

# Atlantis—Growth and Disaster

She took no thought of her dream,
therefore her Fall is terrible.
Jeremiah

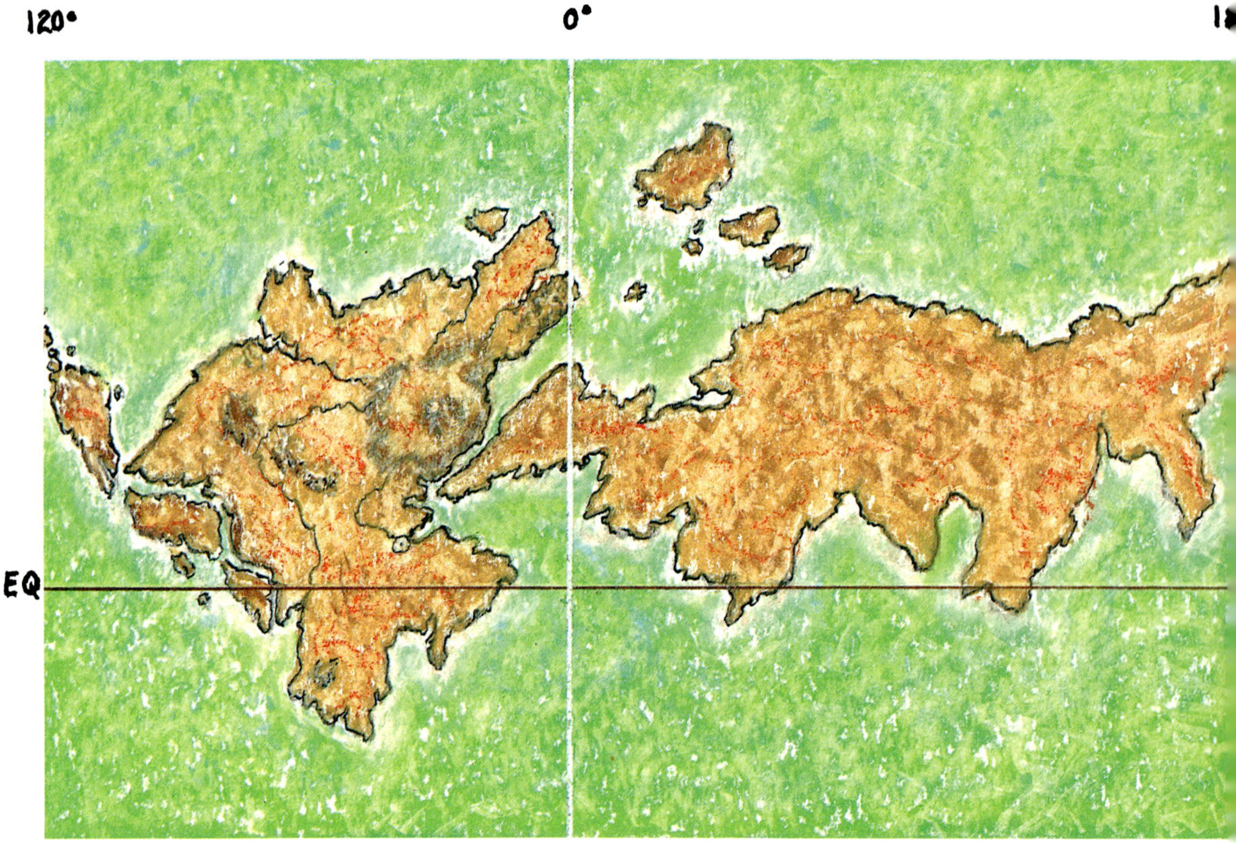
120°
0°
EQ

MEDITATION: In Atlantis, darkest materiality and noblest spiritual wisdom were joined in total struggle for the evolving divine Self in Man. The struggle still exists, in us, though in a different balance.

VISUALIZATION: When fragmented Lemuria was no more—and there remained only the island continent that is today Australia and a lesser one, Madagascar—another vast body of land had risen that stretched across the Northern Hemisphere. This was giant Atlantis, battleground for the warring elements in humanity—the animal and the spiritual. The visualization shows Atlantis at its heights—a land mass in the Northern Hemisphere that almost encircled the globe. In Atlantis the races of mankind evolved the human physical form to near perfection, its emotional counterpart being the arena of ceaseless embattlement.

In the Atlanteans awakened an imperial driving power to explore and colonize and spread their rule and cultural developments abroad in the world. Under this influence, and with the aid of the great Teachers, they created "cities of the golden gates" that became centers of progressive civilization and high attainments in science, engineering, and the arts. They developed majestic forms of government accompanied by religious splendors that brought order in their kingdoms. Mystical rites performed in their temples taught symbolically knowledge of the universal principles that govern all life, together with the reality of Man's inner Self and hidden laws of Nature. The Atlantean initiations into occult science awakened command of invisible realms and forces. This learning made possible the raising of gigantic structures, particularly temples and monuments, some of which exist to this day in various parts of the world.

But these heights of cultural and material achievement were eclipsed by the depths of degradation to which the Atlantean nature was prone to sink when, with the departure of the divine Kings and Teachers, the cycles of enlightened progress closed. Power then fell into the hands of those who perverted it to selfish exploitation. The sacred occult arts were turned to a diabolical subjection of Man to his yet strong animal nature. These practices quickly spread and engulfed the populace in dark iniquity.

The downward course led inevitably to the overturn of Nature's balances and therefore to readjustment through natural cataclysms. In cycles that rose and fell, fresh progress was followed by colossal failures, each shattering further the Atlantean continent. Yet it was the war-ground upon which humanity's future course was decided, even though the victory was not yet universal. That triumph awaits the arrival of a new race upon yet another continent to clear the path of ancient karmic burdens and to open the way into the golden ages of future mankind.

# 26

## The Material Sheaths

Behold in the invisible mansions of our being the self-created progeny that hasten our steps to glory or gloom.

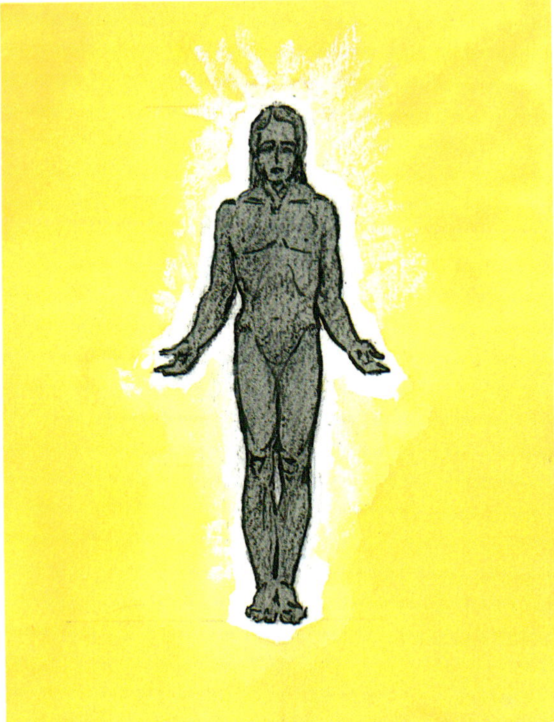

A

B

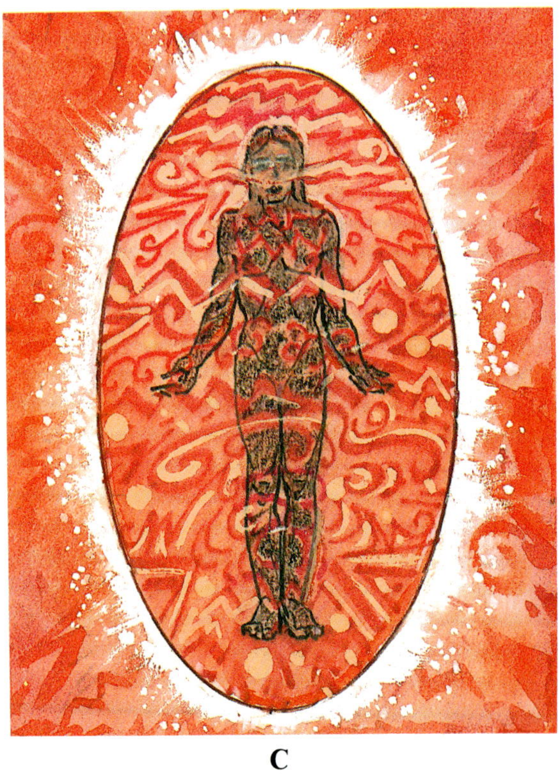

C

D

MEDITATION: With the Atlantean karma affecting the modern environment, Man continues the struggle for spiritual mastery. His four levels of embodiment are in ceaseless motion. His use of them is continuously changing their texture and quality. All four of the bodies—one material, three superphysical—perish at the end of each life span. But their quality and aptitudes live on to reappear in the assembling of the new vehicles of the succeeding incarnation. It can be readily seen that this is vital to living a more harmonious and intelligently organized existence within one's vehicles. Supreme wholeness depends upon informed, responsible usage of the transient bodies upon which our future spiritual stature depends.

VISUALIZATION: Every human being is enclosed in four levels of materiality. The visible physical body is the densest; it is the strong foundation upon which the structure of consciousness rises. The etheric sheath that enfolds every particle of the physical body is of such importance to Man's well-being that it may be useful to consider it a second vehicle. The third is the astral body; and the fourth, the mental body (the denser levels of Manas). Each of these four vehicles exists in a different medium and environment, yet they function together with an incredible harmony of being when all is well. Consider the character of these separate coordinates in the total order:

(A) The *Physical* body regularly demands proper nourishment, sensible exercise, disciplined habits, and a rhythm of adequate rest and activity for its harmonious functioning.

(B) The *Etheric* sheath is the finely sensitive means of transmitting all forces and impulses from the physical brain, nerves, and organs to and from the astral and mental levels of the Self. It also receives and distributes to the physical body the invisible universal forces and nourishment (such as prana from the Sun). Also, this vehicle is a beneficent magnetic shield that protects one's private sphere of focused physical being from open awareness of the occult worlds within and around us.

(C) The *Astral* body is ever churning with our feelings, desires, and emotions. We are projecting these into the surrounding astral world, and thereby attracting and receiving from it impacts of a similar nature. Obviously, we are blessed or plagued by the quality of feelings and desires that we permit entrance into our astral domain. It is clear that the healthy or sick condition of one's emotional self depends entirely upon what one's attention dwells upon habitually—one's astral way of life. The requisite for a healthy astral body is a deliberate association of feelings and emotions with the unified reality: the Good, the True, and the Beautiful. We do this by continuous discrimination among the moment-to-moment choices that are pressed upon us. We can maintain a restful heart in which to retreat, or we can suffer the torments from which there is no escape, because our astral environment rides with us everywhere we go.

(D) The *mental* body is comprised of the lower sublevels of the mental plane and is oriented to material being. We are so closely identified with this aspect of ourselves that usually it is mistaken to be the essential self. Since it perishes at the end of a lifetime, many have a fixed belief that "death ends all." Hence the climate of fear.

It is supremely important to realize that every exertion of lower mental matter creates a form—a thought-form—and all of these forms are polarized to materiality. We are therefore surrounded by the forms that we ourselves are creating. Moreover, our creations attract similar forms. The surrounding mental world is crowded with insipid clouds of mentation mostly concerning current physical needs or desires and their attainment. There is much that is idly viscious, salacious, or cruelly selfish. There are also dangerous, evil, violent, and demoniacal thought-forms to be carefully avoided. On the other hand, there are exalted beneficent and gloriously beautiful forms that are as golden clouds in a serene sky. Any of these will impinge upon and enter our mental body if we attract them. The mental body can also be a bastion against evil forces, an impregnable shelter impervious to surrounding disharmony, if we will.

Particularly useful is the knowledge that whatever quality is ejected from the mental body invites the opposite quality into the vehicle: evil thoughts rejected automatically attract the good, and vice versa. Thus the very structure of the mental body can be gradually transformed and lifted permanently to a higher level that invites the downpouring of luminous and joyous forces from beyond the mind.

# 27

# The Great Divide in Man

Time and space are conquered, and their barriers no longer exist for the unified man.

Annie Besant

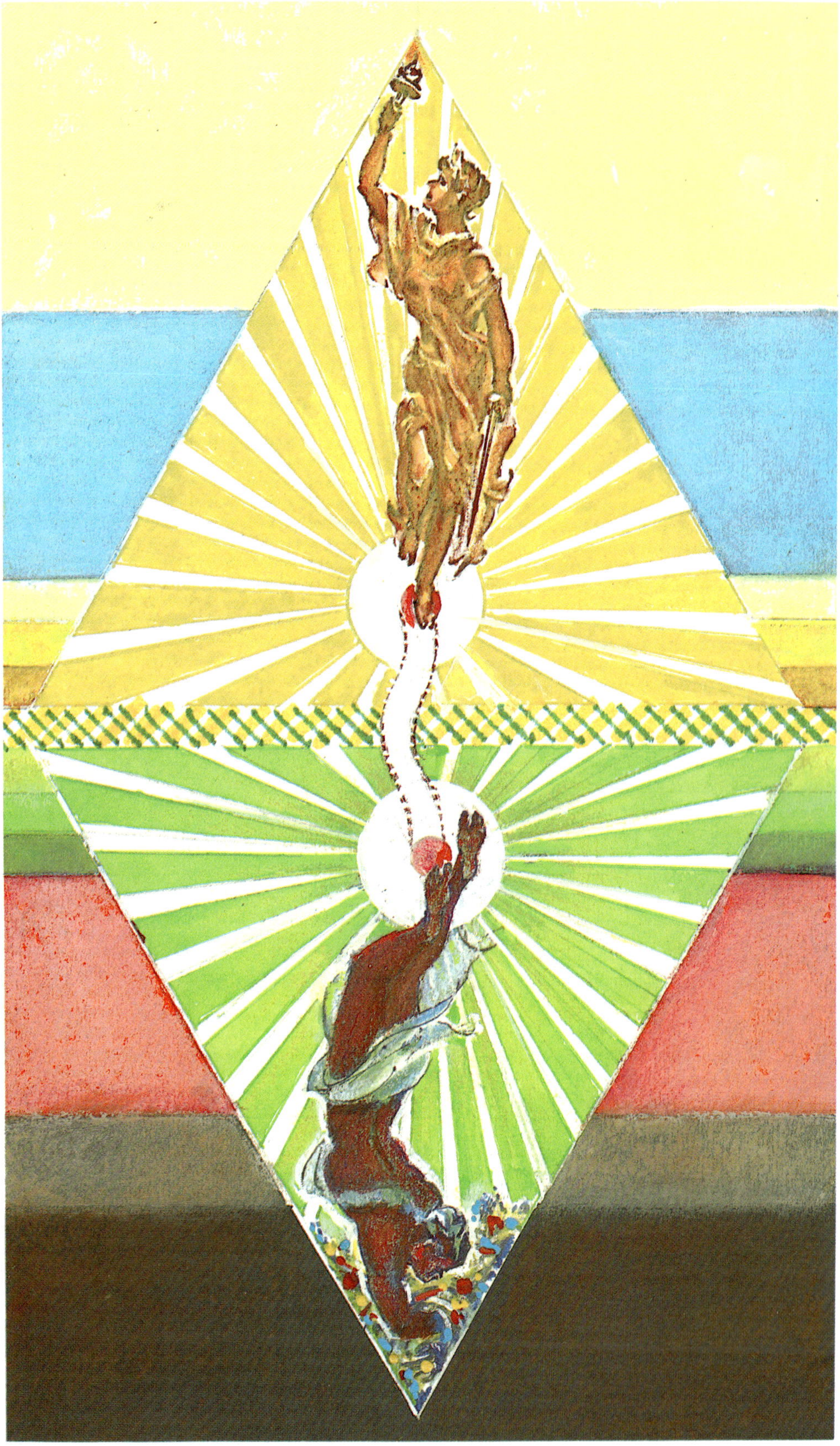

MEDITATION: Midway in the great forests of the mind there exists an invisible barrier more formidable than walls of steel, yet subtler than the unseen prana in the air we breathe. Mere thoughts alone have not the power to bear consciousness across this barricade, while yet thoughts are oriented to materiality.

VISUALIZATION: Envisioned here is the dual-natured, mortal man and his immortal soul. The lower self is shown enraptured with materiality, absorbed in gathering the harvests of Life's possessions. He is delighted with its sensations and transient pleasures, unaware of, or totally inattentive to, a heavenly counterpart. The soul is pictured above reaching upward with the chalice of aspiration to receive the divine power and inspiration pouring downward from the Monad. Yet these two aspects are one Self, even when not consciously unified.

Any energizing of matter in the lower subplanes of the mental body, as said, creates thought-forms or waves that are polarized to matter. All such thinking is directed to physical concerns and material affairs.

On the other hand, any energizing of the higher Manasic mind—where attention is directed to abstract metaphysical, poetic, philosophical, or spiritual thinking—is polarized to life far beyond the mind. Such abstract effort bears consciousness across the ''great divide'' on wings that can reach the loftiest realms and awaken experience of the sublime Self that is beyond all limitation. The seeker who cultivates a discriminating use of the higher mind may gain intuitive experience of the reality that lies beyond the mind.

The seven levels of the mind are divided at the fourth level. The three lower ones are polarized to matter and are classified as the mental body. The three higher ones are opposingly polarized to spiritual heights. These levels are classified as Manas. The fourth level is an intermixture of both polarities. This is the barricade where the transition can take place between the higher and lower levels.

Two circles representing the centers of consciousness in the higher and lower selves are linked together by a passageway that has long been recognized in the traditional wisdom as the *antahkarana.* Through it one crosses unconsciously to the higher Self in exalted and intuitive moments of inspiration and vision.

The triumph will come when one functions harmoniously with unbroken consciousness in mastery over all the levels of the self. Even now we are the same person whatever the level with which we are presently occupied. Those who are recognized as Masters of the Wisdom are in full waking consciousness in which all vehicles are excellent tools for use by the higher Will.

The vehicles are no longer permitted to imprison the Self that has achieved liberation from materiality. The greater Teachers of mankind have reached this freedom; to it we may likewise aspire and devote our energies.

# 28

## Uniting of Consciousness

Veil after veil will lift—but there must be veil upon veil behind.

Sir Edwin Arnold

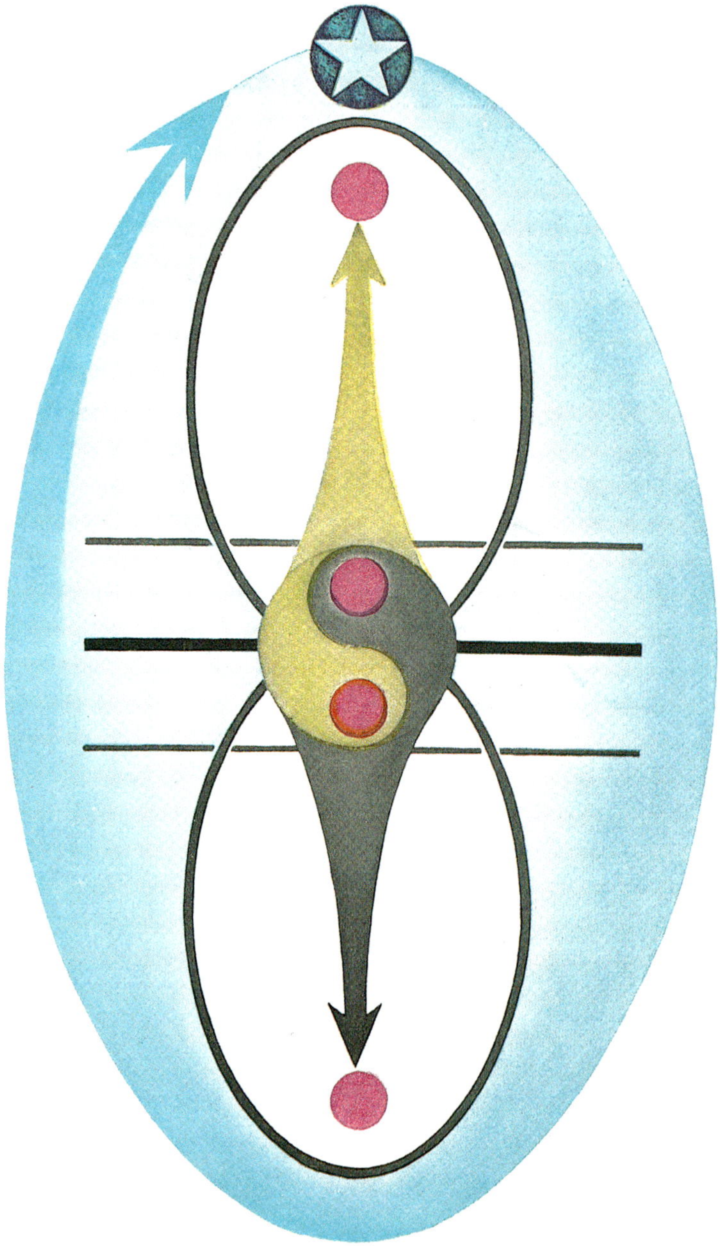

MEDITATION: Nothing exists that is not a part of the "absolute abstract motion" which appeared with the creation of the universe. When the One Consciousness became an infinitude of units, the units themselves continued the ceaseless motion on their various levels, as they became involved in matter and evolved out of it through Nature's kingdoms to humanhood and beyond.

VISUALIZATION: In this absolute motion, there exist seven quantum jumps inward from the physical plane's greatest velocity, the speed of light. No physical object, it is said, can move at velocities beyond the speed of light, for in doing so the object would fly apart atomically and disappear. But such disappearance, viewed occultly, would simply be an object's vanishing into the astral light, of much greater velocity. At each of the seven levels inward, the jump increases toward "absolute abstract motion."

There are multiples of the seven levels in the wondrous complications of Man's consciousness. Harmonizing these into united being is a high aim of learning and wisdom. Is there, then, some basic pattern in the movements of human consciousness—some governing principle that offers a key to the means of control?

If you observe a movement of conscious attention aroused by a stimulus, say at the physical level, you may note that it is immediately amplified by attraction or repulsion (the astral level) and becomes further energized by visualizing its fulfillment or rejection (the mental level). This stirs the feelings compellingly (again astral), and the cycle of motion ends in the succeeding physical action; or it is renewed through frustration. The pattern moves through the three levels of mortal man's being: physical, astral, and mental, and returns to equilibrium in its source. This suggests a principle of motion, elliptical in nature, a movement around two opposite centers or foci—one located at the physical level, the other at the mental.

One may see here an incessant pursuit of negative fields by positive forces. The positive force automatically seeks the negative field. If the positive is physical, the negative will be astral. Reaching there, the force is transformed by identity with the field into a negative force seeking the next positive (mental) pole. Transformation again takes place there rendering the force once more positive, moving it around the pole and, via the astral field, back to physical fulfillment, obeying the elliptical principle.

A statement of the principle, therefore, is "from positive pole to positive pole across a negative field." An example: hunger begins in the physical body; this arouses astral desires; these stir mental visualization, which whips compelling astral desire for action. The cycle reaches equilibrium or frustration in the resultant action. The principle has endless applications.

Such orbits tend to become fixations. The elliptical motion hardens into a spherical equilibrium—a "nongrowth" position of inertia. Such inert cycles tend to remain fixed, yet are temporary because of life's impacts from without, such as calamitous effects from adverse karmic blows or from implacable urges that arise from within. These upset the nongrowth equilibrium. The consequent high intensity of anxiety will force consciousness painfully to find a new orbit.

However, a new orbit of spiritual growth can be attained in other less harassing ways. Orbital meditation is a deliberate use of the orbital principle to intensify consciousness through spiritual aspiration that reaches beyond the mental to higher intuitive levels of experience. Through meditation the elliptical pole in the lower mental level can be lifted to become in Manas a new lower pole of the Manasic-Buddhic-Atmic orbit (as shown in Visual 28). With persistent meditation, accompanied by a growing purity of life, the higher orbit can be opened and become more firmly established. A repolarization of the Self can take place. A new spherical equilibrium is reached—a resting place in immortal consciousness, beyond the wear and tear of slowed-down mortal life. With the new orbit, time barriers vanish in a dawning sense of *cosmic leisure.* The pressures of mortal existence are eased by a growing spiritual tranquillity.

The accompanying visual design illustrates this action. The mortal and immortal orbits are interwoven, the higher bringing down vision into the lower mind and the lower lifted by the higher mind. The whole is oriented to the highest spiritual level—the star—representing the higher Self.

As the illustration suggests, perfection is motivated from the *Atmic* level—the spiritual Will. The motive power from there drives downward into physical action and returns automatically to its source above. Adepts in this spiritual art of mastering the geometry of movements in consciousness can unite both mortal and immortal elements in a full and fearless consciousness of wholeness. They walk the streets of earth as immortals, knowing the fruits of wisdom.

# 29

# Reincarnation—Gradual Spiritual Growth

. . . a thousand times slain, but not once killed.

H. Munro

MEDITATION: We have viewed vast perspectives of that fragment of the Supreme One—the Monad—whose consciousness descended into deepest materiality, and there evolved the individualized soul in preparation for the cyclic return of the Self, as united body, soul, and spirit, to its source in God. The cyclic return is through the process of birth-death-rebirth in physical bodies. In each incarnation latent powers in the soul awake and are developed.

VISUALIZATION: The mighty rhythm of reincarnation repeatedly draws the soul into the maelstrom of physical life in the various historical periods, races, and civilizations of the world. Defeats and triumphs are experienced, challenging the soul to create ever greater mansions of Life's material substance. At the close of each lifetime on earth the Self returns "home" to prepare for a new day.

The great cycle of normal reincarnation thus begins in the remote regions of the immortal soul. The descent takes place into the mental and astral worlds, where the permanent atoms are activated, gathering around them the materials of new mental and astral vehicles generated by their previous condition. Nothing of the past is lost. Then follows the forming of an etheric vehicle, or matrix, within which the dense physical body is shaped by the forces of pending karma, modified by the soul's destined opportunities in the new life. Parentage, race, and nationality are predetermined by past relationships—good or ill—and by service to one another.

The universal rhythm of reincarnation cannot be fully comprehended without relating it with the accompanying principle of balance (karma), which is the bridge between incarnations. Always and everywhere this principle is functioning in endless chains of action-reaction. Every exertion of energy—mental, emotional, or physical—becomes a cause that produces its results. A knowledge of the operation of this universal law throws great light on the subject of reincarnation and the overall purpose of life.

As shown here, the process of rebirth actually begins with the soul's departure from heaven-life, in which a vision was beheld of the new incarnation's favorable circumstances for growth, as well as karmic limitations. This glimpse constitutes a preview of 'soul-duty' to be observed if there is to be inner harmony in the life ahead. With the mental and astral bodies assembled, the etheric pattern that determines the physical body's structure and condition is formed in the mother's womb. Into this shielded place the soul descends and in the new brain loses all memory of the past.

A new life on earth begins normally in the pure innocence and natural happiness of the infant, protected by parental love and upbringing. Next we see the youngster meeting his new life full of zest and excited anticipation. The years of adulthood bring great trials of suffering and struggle as well as achievement. Finally in old age the incarnation ends; the outworn physical body is dropped and disappears forever.

The astral sojourn is simply a period of disentanglement from material attachments. The Self moves upward in its growing freedom from the recent limitations. The heaven-life is a joyous fulfillment of all relationships, all unrequited hopes, loves, and dreams wished for on earth. All are brought to happiest fruition and peaceful harmony. Finally there is the blessed uniting of soul with divine spirit in wondrous concord. If the element of fear is removed, falling asleep and peacefully dying are similar experiences. If we rejoice in the refreshment of a restful night's sleep, we need not fear death.

Many centuries are required for this inner segment of an incarnation's cycle. Immediate reincarnations are abnormal and bear no witness to the reality of this greater cycle in the soul's gradual unfoldment through the aeons.

# 30

## Meditation—Door to Swifter Unfoldment

The soul of Man is immortal. . . its growth
and splendor has no limit.

*Idyll of the White Lotus*

A

B

C

D

MEDITATION: Life eternally moves from imprisonments through death of the old to new freedoms and rebirth in the new. Each death entails a letting go of what had been fiercely held as essential to one's very existence. The chastening experience of "letting go" purifies the self from material entanglements through desirelessness.

The goal of inner freedom need not await this slow, painful growth toward a far-off future attainment; it can be realized at present through meditation that repeatedly opens the door to the experience by a simple observance of the knowledge of the Self, as indicated in this book.

VISUALIZATION: First, one is seated quietly and comfortably, becoming gently alert to the sphere of space that surrounds each of us (Visual A). The individual is king within his own space sphere, or field. In time he becomes its ruler at all levels. Now try to realize the vital prana that is radiating from the sun, nourishing all living creatures (Visual B). With a deep breath, using the will, draw in that essence, visualizing it as reaching the deep center of yourself, as a radiant healing and energizing force. Think of the center as near the heart region—*never* below. Hold the breath momentarily while letting the full radiance of the prana reach every part of the physical body: head, trunk, limbs, hands and feet. Now exhale slowly, intentionally breathing away all disharmony, all disturbance and minor pain. In the stillness let peace radiate from within outwardly to the surrounding environment. You are beginning the movement of consciousness in its lower orbit by rendering the physical center positive.

Now lift full awareness to the larger astral sphere of feelings and emotions (Visual C). Visualize this as interpenetrating the physical sphere. Consciousness, seated in the physical and etheric vehicles, is now focused in the astral body. Observe it generating feelings and desires. When these clash tumultuously, the whole astral body becomes a stormy sea of forces. But that sea is being rendered calm and serene by your will, which is situated at the center, ruling both the physical and astral bodies. Again, breathe in the vital *astral prana,* which is shining there even more brilliantly than in the physical world. Try to experience a deep, pure, selfless and impersonal love irradiating the whole astral body with the inbreathing. Hold the breath steadily, experiencing the hidden beauty and joy that accompanies this selfless love. As you exhale, breathe away all ill-feeling, all ugliness and unhappiness. Now with the will visualize your whole astral domain as a veritable lighthouse of peace shining in the storm-tossed painful and confused astral world. Let that peace radiate to all beings.

Now lift the orbit to the larger mental sphere (Visual D) that envelops both the astral and physical spheres, and observe the self in this greater environment. The mental sphere presents an even more formidable challenge to the inner ruling Self as it is in ceaseless motion. The problem is to bring about stillness in this constantly churning region. Once more using the breathing technique, draw in the brilliantly radiant *mental prana,* holding it steadily focused in the mental center, while sending the prana shining throughout the mental sphere. Then slowly exhale, breathing away all material thought-forms, the mental body remaining filled with Light only. Repeat until the mind remains perfectly still, poised, and alert, reflecting steadily the Light of the higher Self (symbolized by the Star). The orbit has reached its highest point, and will move automatically back to its physical source with resultant action or frustration; or with higher energies awakened, the gate may be opened to the greater orbit beyond. Awareness of the higher Self may appear at first as a faint light shining in one's inner horizons. The light becomes brighter and the sense of direction surer with experience. Try to realize the Immortal Self more firmly. When the Self begins to take hold, the path to the heights will be assured.

By this controlled orbital movement one may not only bring about a tranquil, harmonious state in the three lower levels, but the mortal person will be rendered a radiant center of higher forces. The Light within can now open a passage for the consciousness to cross the threshold to the higher orbit that is oriented to the pole opposite from materiality.

# 31

## Passage to Freedom

He lives now not in the world, but with it;
His horizon has extended itself to the width of
the whole universe.

*Light on the Path*

E

F

G

H

MEDITATION: The secret of activating the higher mental sphere—Manas—lies in the fact that no binding material thought-forms are created there. The Manasic energies are directed to subtler levels. Truth is sought through abstract higher reason, intellect, and intuition. The pursuit therefore must go beyond the mind into Buddhic and Atmic levels. The movement of consciousness into a higher orbit is a tremendous step into transcendental regions where the mortal outlook is expanded to universal horizons.

VISUALIZATION: In this more advanced meditation think of the Self at the center as before—not only centered in the physical, astral, and mental spheres, but in the higher mental as well (Visual E). This can be realized by deliberately lifting consciousness to the Manasic level. Try silently chanting some sacred verse; or envisioning vividly the beauty of some inspiring art or architectural creation; or listen imaginatively to glorious lines of poetry; or hear inwardly a passage of music that floods the whole self with joy and wondrous upliftment. In many ways one's consciousness may be lifted to a rapport with the immortal Self.

With the higher orbit of consciousness opened in this way, the elliptical motion commences (Visual F). Consciousness mounts toward the pole opposite to materiality. The movement goes swiftly—even instantly—to the Buddhic field where all perceptions become illumined by pure love and inspired intuition. A profound union with Nature and its kingdoms can be experienced, as oneness with the lands and the mountains, the continents and oceans, the heavens and the whole global earth, and space beyond—in a creative union of the whole. If this vastness seems overwhelming, it is because Divine love has not yet been realized adequately. Divine love is the precious essence that, when more fully awakened, lifts consciousness beyond the Buddhic to the Atmic level in a veritable baptismal rebirth into nirvanic bliss (Visual G). Beyond is an even higher, Divine orbit.

In it the great Star (Visual H) represents the *original Self*—the Monad, a spark of the Logos. It is one's Archetypal Being—the pattern of perfection toward which Life is ever evolving in one. It glows with the will to truth, to beauty, to harmony and fulfillment. One who enters this sublime state of union and of peace that is power in equilibrium is able to perceive universal archetypes more clearly and to unveil profounder aspects of Reality within oneself.

At the Atmic level one stands in the presence of one's true Teacher, having drawn nearer to Him through "a purified heart and steadily strengthening will." A revolutionary change has taken place both in outlook and attitude toward life. Here begins the disciplined life that leads to liberation. The higher Self takes command of the lower as indicated in Visual 28, which illustrates the mortal orbit firmly joined with the immortal. At this point, as shown, *all* action is motivated from the Atmic level through what is perceived to be true action. Although the original impetus reaches downward to physical expression, the orbit of consciousness is not completed there; it sweeps back upward to its source in Atma. These are moments experienced in immortal wakefulness. Each precious one illumines and further refines the lower bodies. With attainment of such a spiritual level, the soul rests in a peace sublime. From that condition of spiritual leisure one may begin to explore the greater horizons of one's divinity.

The fully awakened soul becomes a beneficent force in the world, joyfully ready to serve wherever located in the turmoil of daily life. Such a one labors to change the darker material of earthly life into finer textures that will more nearly accommodate the archetypes as projected in the beginning. So awakened, the individual is an unclouded Light illumining the path to freedom for others.

Life is ever groping for new paths to excellence. In Man this urge is the yearning search for *spiritual freedom.* Even though physical and mental freedoms may be secured, one remains in spiritual bondage. The ultimate freedom-marchers are those who through illumined knowledge, having perceived a practice that leads to union with the Eternal Whole, apply themselves diligently to their path of return.

As our visualized meditations arrive at this point, they reach their natural conclusion.